YOUR KNOWLEDGE HAS VALUE

- We will publish your bachelor's and
  master's thesis, essays and papers

- Your own eBook and book -
  sold worldwide in all relevant shops

- Earn money with each sale

Upload your text at www.GRIN.com
and publish for free

Cosmic Ray Intensity Variation During the Passage of Interplanetary Disturbances

Rajesh Kumar Mishra

Rekha Agarwal

Bibliographic information published by the German National Library:

The German National Library lists this publication in the National Bibliography; detailed bibliographic data are available on the Internet at http://dnb.dnb.de.

ISBN: 9783964879660
This book is also available as an ebook.

Print and binding: Books on Demand GmbH, Norderstedt, Germany
Printed on acid-free paper from responsible sources.

The present work has been carefully prepared. Nevertheless, authors and publishers do not incur liability for the correctness of information, notes, links and advice as well as any printing errors.

GRIN web shop: https://www.grin.com/document/1443130

COSMIC RAY INTENSITY VARIATION DURING THE PASSAGE OF INTERPLANETARY DISTURBANCES

Rajesh Kumar Mishra[1] and Rekha Agarwal[2]

[1] Information Technology Cell, ICFRE-Tropical Forest Research Institute

P.O.: RFRC, Mandla Road, Jabalpur (M.P.) 482 021, India

[2]Department of Physics, Govt. Model Science College (Autonomous)

Jabalpur (M.P.) 482 001, India

Abstract

An investigation has been made so as to study the cosmic-ray decreases occurring during 2006 with respect to the arrival times of interplanetary shocks and magnetic clouds. We have identified three interplanetary magnetic cloud events during 5, February 2006, 13 April 2006 and 14 April 2006. The interplanetary magnetic field (B), north south component of interplanetary magnetic field (Bz), solar wind velocity, sunspot number (R) and disturbance storm time index (Dst) associated with these events has been studied in the present work. The data (neutron monitor count rate) from Newark Neutron Monitor 9NM64 has been used. The north south component of IMF (Bz) produce large geomagnetic disturbance on the onset of interplanetary magnetic clouds. The deviations in the interplanetary and solar wind plasma parameters are significantly correlated to the magnetic cloud events. The increase in Dst index, sunspot number (R) and Bz after the magnetic cloud event produces increase in cosmic ray intensity.

Introduction

Interplanetary Magnetic clouds :

Magnetic clouds belong to one of the several classes of transient flows in the solar wind. Magnetic clouds as ideal force free objects (cylinders or spheres) are ejected near the sun and followed beyond the Earths orbit. Magnetic clouds were discovered in 1970 by in situ space craft measurements as region in the solar wind where (1) the magnetic field magnitude is higher than the average: (2) the smooth magnetic field vector rotates through a large angle order of a day (3) proton temperature is lower than the average (Burlaga, 1988). All three of these criteria must be satisfied if an event is to be identified as a magnetic cloud. A cylindrical magnetic cloud is introduces in the computational region through the inner boundary (at 18 solar radii) as a Lundquist force free solution (Burlaga, 1988). This cloud has its axis lying in the ecliptic plane. Another spherical magnetic cloud is also introduce in the same inner boundary (at 18 solar radii) as a Chandra Sekhar force free solution (Vandas et al., 1993 a).

The magnetic field configuration in a magnetic cloud is approximately force free (Goldstein, 1983). A magnetic cloud is called a positive of interplanetary magnetic field is directed north ward at the time of on set, and a negative cloud if the direction of interplanetary magnetic field is directed south ward. Based on an analysis of Multispace craft data. Burlaga et al (1990) have shown a global view of the field lines speculated for a such structure out of 1AU from the Sun. Clouds have a structure of radial dimension $\sim$ .25 AU at 1AU. The field geometry in a magnetic could is consistant with a magnetic loop. In the cloud, the field strength is high and density and temperature are relatively low. Thus the dominant pressure in the cloud is that of the magnetic field. The total pressure inside the cloud is higher than outside, implying that the cloud is expanding as it moved out ward, even distance of 2AU (Burlaga et al. 1981). Its geometry is that of a helical field lines confined to a flux tube, which is covered on the large scale i.e. a curved flux rope (Farrugia et al, 1997). The Pitch angle of the helical field lines increases with increasing distance from the axis of the magnetic cloud, such the field lines aligned with the axis of symmetry at the position of the axis and perpendicular to it on the clouds boundary. The magnetic clouds can interact with other flows (Burlaga, 1995). Magnetic clouds are ideal objects for solar terrestrial studies because of their simplicity and extended intervals of southward and northward magnetic fields (Burlaga et al, 1990). Gosling (1990) has shown that approximately 1/3 of interplanetary manifestations of solar ejecta (also called CMEs by some authors) are magnetic clouds. A magnetic cloud like a well defined CME acts as a driver and forms a driver shock wave; propagation and properties of the shock and driver it self. Hence like interplanetary shock, magnetic clouds provide us with a link between ejected material, field and energy on the sun and significant magneto spheric activity via solar wind. Many workers have shown the structure and dynamics of interplanetary magnetic clouds and their effects on the magneto sheath and magnetosphere. Reviews of the association of geomagnetic activity to magnetic clouds and other IMF features are given by Farrugia et al. (1997) and

Tsurutani and Gonzalez (1997). Farrugia et al (1996) have shown that a major geomagnetic storm and associated aurora were produced by the extended interval of the negative B_z in the front part of magnetic cloud. As the magnetic cloud moved past the Earth, the magnetic field slowly rotated northward giving an extended interval with positive B_z in which the geomagnetic activity subsided.

Nature of Magnetic clouds :

Cocconi et al. (1958) suggested that the magnetic field lines in a cloud form an extended loop, with ordered fields, the field lines being an chored in the Sun, and they called such a loop, on elongated tongue and a magnetic bottle. A similar concept was discussed more quantitatively by Piddington (1958), who considered the additional possibility that a loop could become detached from the sun, forming closed magnetic field lines in the solar wind (a magnetic bubble). Gold (1959) proposed that the magnetic loop might be preceded by a shock wave. All of these workers envisaged that the magnetic cloud or bubble or loop is formed by motion of plasma ejected from flare or some other solar disturbance. None was very specific about the three dimensional configurations of the magnetic field. High magnetic field created by a shock wave (Parker, 1963); irregularities in the magnetic field (Morrison, 1956; Laster et al., 1962); a series of directional discontinuities (Barden, 1973); large scale tangential discontinuity (Quenby, 1971; Lockwood et al., 1975) a magnetic blob (Barough and Burlaga, 1975, Kane, 1977) and a spiral cone like region which extends along the interplanetary magnetic field lines from the sun to the radially advancing front generated by a type IV solar flare (Iucci et al., 1979 a).

In two independent studies (Burlaga et al., 1981; Sanderson et al., 1983) magnetic field and plasma parameters were used for analysis. Magnetic cloud is defined as a flow system consisting of a shock, a turbulent sheath and a ordered magnetic cloud in transient ejects associated with the shock, using magnetic field and plasma data from fine space crafts. The emphasis is on the magnetic cloud which was identified by a characteristic variation of the latitude angle of the magnetic field. As the magnetic cloud moved past each of the spacecraft the magnetic field direction was observed to change by rotating nearly parallel to a plane. Thus the magnetic field configuration in the cloud was essentially two dimensional. The orientation of this plane of maximum variance with respect to the space craft, sun line and solar equatorial plane was the same at all of the space craft with in $\sim 20^0$. These results suggest that the lines of force in the magnetic cloud formed loops, but it could not be determined whether these loops were open or closed. Inside the magnetic cloud the speed was high and the density and temperature were relatively low, especially near the middle of the cloud. The total pressure in the cloud was higher than the ambient pressure at 2AU, indicating that the cloud was probably expanding at 2AU and by inference, with in 2AU as well. The magnetic pressure was larger than the thermal (ion plus electron) pressure indicating that the expansion was derived primarily by the magnetic field which presumably originated in some transient process at the sun. Expansion driven by the high magnetic field pressure inside the cloud is at least one cause of the low density and temperature in the magnetic cloud, the input conditions might be another cause.

The momentum flux in the cloud at 2AU was not generally higher than that a head of the cloud, yet the cloud was preceded by a shock. It is suggested that the shock might have been driven by the stream carrying the magnetic cloud when it was near the Sun, but that the momentum flux decreased in transient no. 1AU owing to expansion and perhaps deceleration so that at 2AU the shock was no larger driven but rather moved on a head of the cloud by virtue of the motion it acquired earlier of course we can not exclude the possibility that the shock and ejecta were created indepently at the sun. At the near of the magnetic cloud. There was most unusual filament characterized by high proton density n and low B and T, with very thin boundaries having the nature of tangential discontinuities. This filament was in equilibrium with the medium in which it was embadded. Its boundaries (current sheet) were nearly parallel to the plane of maximum variance of B in the magnetic cloud. Current sheets in the sheath a head of the magnetic cloud had different orientations more nearly perpendicular to the radial direction and parallel to the surface of the shock wave.

Size of the Magnetic clouds :

Two significant conclusions about the size of the magnetic cloud are : the azimuthal extend of the magnetic cloud is $\geq 30^0$. Nothig is known about the latitudinal extend of magnetic clouds. The radial dimension of clouds observed at ~ 1AU is typically ~ 0.25 AU (Klein and Burlaga, 1982).

Type of Magnetic clouds :

Magnetic clouds can be classified on the basis of their relations to the plasma parameters, eg, temperature, density, speed, magnetic field intensity etc. There are three types of magnetic clouds as follows.

(i)   Clouds preceded by shock (SAC)

(ii)   Clouds followed by stream interface (SI)

(iii)   Clouds associated with cold magnetic enhancement (CME)

CME is a region in which the plasma temperature is anomalously low and the magnetic field anomalously low and the magnetic field strength is enhanced (Burlaga et al., 1978). This classifications is similar to that by Burlaga and King (1979) for enhancement of magnetic field strength. The field and plasma parameter in each classes are similar suggesting that the three types of clouds might be different manifestations of single phenomena (e.g. coronal transients).

Interface clouds may have been swept up by corotating streams. Shock associated clouds more faster than the other two types, which are basically slow flows. The magnetic pressure inside the clouds is higher than the ion pressure of the material outside of the clouds.

Influence of magnetic clouds on comic ray intensity:

Magnetic clouds preceded by shock and cosmic ray intensity variations

It is found that the decrease in cosmic ray intensity, which are associated with magnetic cloud preceded by a shock, are very high and these decrease starts few days earlier than the arrival of cloud at Earth. From the study of the time profile of these decrease, it is found that the onset time of a forbush type decrease produced by a shock associated cloud starts nearly at the time of arrival of the shock front at the Earth (Duggal et al., 1983; Badruddin et al., 1986) and the recovery is almost complete with in a week. Forbush decreases associated with shock associated cloud are caused by magnetic field variations associated with interplanetary disturbances (Badruddin et al., 1986). Whenever, the shock arrival time is not available, sudden commencement of geomagnetic storm (SSC) data is used, since SSC can be regarded as the geomagnetic signatures of the arrival of the interplanetary shock (Burlaga and Ogilvie, 1969). Generally the Forbush type decrease starts at the time of arrival of shock wave at Earth. However, there are a few cases when no appreciable decrease in cosmic ray intensity is observed with the shock associated cloud. Suggestion have been made in the past that the Forbush type decrease is caused by the entry of the Earth in to loop or tongue of interplanetary magnetic field lines that are freshly ejected from the sun. Burlaga et al. (1981) have noted a decrease in comic ray intensity, beginning with the arrival of a shock, but with the minimum intensity occurring before the arrival of the magnetic cloud. Burlaga et al. (1982) have identified a magnetic cloud that could be associated with a coronal mass ejection. Also Burlaga et al (1985) have found that the CR intensity did not decrease further during the passage of the associated magnetic cloud. On the other hand, Badruddin et al. (1985) have reported a possible correlation between magnetic clouds and cosmic ray intensity decrease while Kudo et al (1985) have reported an increases in cosmic ray intensity that may be related to the geomagnetic D_{st} index and Iucci et al. (1985) have found short term increase in CR intensity occurring inside the Forbush decrease, that possible may be associated with magnetic clouds. The total pressure is high inside the magnetic cloud. Due to this fact that the magnetic cloud would tend to expand in order to reduce the pressure. Expansion would produce low temperature and density, and it would reduce momentum flux thereby causing a deceleration of the stream. This concept is essentially a hypothesis which can only be tested by simultaneous measurements by several space crafts (Burlaga et. al., 1984). Thus as a conjecture it emerges out of this discussion that either the shock is produced closer to the sun because of the differences in the momentum flux of the magnetic cloud and the ambient solar wind blowing a head of it, or the shock and the cloud are emerges out of some transient changes (e.g. solar flare, coronal mass ejection etc.) which have taken place at the solar disc; or the origin for shock and the cloud are different transients occurred at the solar disk it self. The decrease in cosmic ray intensity occurs primarily during the passage of the sheath between the shock and the magnetic cloud. Since the rms of the magnetic field is relatively high in the sheath and relatively low in the magnetic cloud. Zhang and Burlaga (1988) infer that the cosmic rays are mainly modulated by fluctuation rather than by drifting in the strong smooth field in the magnetic cloud. Recovery of the cosmic ray intensity for the events associated with shocks is seen to begin during the passage of the magnetic cloud. The shock associated clouds appear to be moving faster than the ambient solar wind a head, where as the other types of cloud (SI, CMEs) are not, possible this relative motions is driving the shock.

Magnetic clouds followed by stream Interface and cosmic ray Intensity variations :

A cloud associated with stream interface is identified with the simultaneous increase in temperature (T), increase in speed (V) and decrease in density (n) at a time when magnetic field strength (F) is maximum. The magnetic cloud is identified with the 24 hour interval in which the magnetic field changes from a northern direction to southern direction and magnetic field strength is high (Klein and Burlaga, 1982).

Periods of enhanced solar wind speed have been associated with coronal holes and active regions, which themselves are characterized by entirely different features. Coronal holes are regions of low density and temperature and occur in a week, open diverging unipolar region, where as active region show a complex magnetic structure including closed field zones evolve rapidly when for ex energetic flares occur in such region. Fast solar wind streams coming from coronal holes and other coming from active regions producing type IV solar flares. Iucci et al (1979b) have shown that during the high speed streams coming from coronal holes the cosmic ray intensity is depressed, the time behaviour of the depression follows the time profile of the wind speed and the stream coming from active region is accompanied by Forbush behaviour are not directly related to the speed increase.

Stream interfaces appear to be a necessary and sufficient condition for a corotating stream. Burlaga et al. (1984) follow Burlaga and King (1979), in using the presence of stream interface an indication of the presence of a corotating stream.

Magnetic cloud associated with cold magnetic enhancement and cosmic ray intensity variation :

In the cloud the temperature is low, the magnetic field strength is high and the speed is near average, these are characteristics of a CME. The boundaries of the cloud are chosen on the basis of the temperature and field strength profiles. The decrease in cosmic ray intensity in relation to the clouds associated with cold magnetic enhancements is smaller in comparison to the decrease in relation to clouds associated with shocks and interaction regions. Moreover, the decrease starts when the clouds associated with CMEs are observed near the Earth. The transient modulation is in this case, not of longer duration. The observed magnetically closed structure in the solar wind containing low temperature, interplanetary plasma are examined in relation to cosmic ray flux examined in relation to cosmic ray flux decrease (Geranios and Rossenbauer, 1983). It is suggested by these authors that small amplitude cosmic ray decrease seem to be related to the observed magnetically closed regions in the solar wind.

Interplanetary magnetic clouds belong to one of the several classes of transient flows in the solar wind. Magnetic clouds as ideal force free objects (cylinders or spheres) are ejected near the sun and followed beyond the Earths orbit. It is found that the decrease in cosmic ray intensity, which are associated with magnetic cloud preceded by a shock, are very high and these decrease starts few days earlier than the arrival of cloud at Earth. From the study of the time profile of these decrease, it is found that the onset time of a Forbush type decrease produced by a shock associated cloud starts nearly at the time of arrival of the shock front at the Earth [1-2] and the recovery is almost complete with in a week.

Nowadays, the analysis of spacecraft data reveals that these events are common in the solar wind. About 30% of coronal mass ejections (CMEs) observed in the solar wind exhibit internal field rotations, characteristic of magnetic flux rope. However, the relationship between the CMEs observed near the Sun and magnetic clouds is poorly understood.

Forbush decreases associated with shock-associated cloud are caused by magnetic field variations associated with interplanetary disturbances [2]. Badruddin et al. [3] have reported a possible correlation between magnetic clouds and cosmic ray intensity decrease while Kudo et al. [4] have reported an increases in cosmic ray intensity that may be related to the geomagnetic D_{st} index and Iucci et al. [5] have found short term increase in CR intensity occurring inside the Forbush decrease, that possible may be associated with magnetic clouds. Zhang and Burlaga [6] infer that the cosmic rays are mainly modulated by fluctuation rather than by drifting in the strong smooth field in the magnetic cloud.

The magnetic clouds can interact with other flows [7]. Magnetic clouds are ideal objects for solar terrestrial studies because of their simplicity and extended intervals of southward and northward magnetic fields [8]. Gosling [9] has shown that approximately 1/3 of interplanetary manifestations of solar ejecta (also called CMEs by some authors) are magnetic clouds. A magnetic cloud like a well defined CME acts as a driver and forms a driver shock wave; propagation and properties of the shock and driver it self. Hence like interplanetary shock, magnetic clouds provide us with a link between ejected material, field and energy on the sun and significant magneto spheric activity via solar wind.
These events are not always associated with interplanetary shocks but only when they travel faster than the ambient solar wind. Besides the identification of magnetic clouds locating cloud boundaries is an open problem Lepping et al., [10]. Zhang and Burlaga [6] showed that the clouds are usually spatially shorter than the interval defined by counter-streaming electrons, suggesting that the clouds are parts of larger transient structures.

Many workers have shown the structure and dynamics of interplanetary magnetic clouds and their effects on the magneto sheath and magnetosphere. The association of geomagnetic activity to magnetic clouds and other IMF features are given by Farrugia et al. [10] and Tsurutani and Gonzalez

[11]. Farrugia et al. [12] have shown that a major geomagnetic storm and associated aurora were produced by the extended interval of the negative B_z in the front part of magnetic cloud. As the magnetic cloud moved past the Earth, the magnetic field slowly rotated northward giving an extended interval with positive B_z in which the geomagnetic activity subsided.

Data and analysis

The temperature and pressure corrected hourly data (counts of neutrons) of cosmic ray intensity from Newark Neutron Monitor (Latitude 39.70N, Longitude 75.70W, Altitude 50 m, Standard pressure 1000 mb, Geomagnetic cut-off rigidity 2.09 GV) have been used, where the long-term change from the data has been removed by the method of trend correction. The days of Forbush decreases have also been removed from the analysis to avoid their influence in cosmic ray variation. Interplanetary magnetic field and solar wind plasma data have been taken from the interplanetary medium data book.

Velocity of solar wind :

The velocities of the solar wind observed so far are given below :

Measured bulk velocities 200-900 Km./sec. Over all average velocities 400-500Km/sec. Average velocities for quite periods 300-500Km/sec (Wilcox and Ness, 1965; Ness, 1967; Hundhausen, 1968b, 1970). The data for Vela 2A and 2B satellites give the mean velocity of the solar wind as 420 Km/sec for quiet periods (Strong et al., 1967). According to observation from Vela 3 satellite this mean velocity is 400 Km/sec and the median velocity is 380 Km/sec for the period from Jul 1965 to Jun 1967 (Hundhausen et al. 1970). These observed values agree well with those, which have been theoretically predicted.

The bulk velocity is normally taken to be very nearly radial component of the solar wind velocity in the ecliptic plane. The long term synoptic measurements at Mt Wilson and Wilcox observatories have been used to verify that in ecliptic solar wind speed is inversely correlated with the rate at which flux tube expand the corona (Sheeley, 1991). The average departure of the wind direction from the ecliptic plane can be with in $\pm5^0$ but some time can go as high as $\pm15+0\ 20^0$. Ulysses mission predict that the solar wind speed was fast and nearly constant above 50^0 latitude. Compositional differences were observed in slow (low latitude) solar wind and in fast (high latitude) solar wind. The radial magnetic field did not change with latitude, implying that polar cap magnetic fields are transported towards the equator. The intensity of galactic cosmic rays was nearly independent of latitude. Their access to the polar region is opposed by out ward travelling, large amplitude waves in the magnetic field (Smith et al., 1995).

Interplanetary magnetic field :

The interplanetary magnetic fields are the result of extension of the lines of force of the general solar magnetic field into space by the expanding chromospher and corona. Thus the interplanetary magnetic field (IMF) B follows from the solar wind velocity V through the hydromagnetic equation.

$$\frac{\partial B}{\partial t} = \nabla \times (V \times B)$$

The source of interplanetary magnetic field (IMF) is the solar magnetic field, inparticular the photosphere field that is swept out by the solar wind. The general magnetic field. On the surface of the Sun is about one gauss. Near the poles, the field looks like a dipole and becomes radial at larger helio centric distances near the equator the field is rather disordered. Since the continuously ejecta solar wind is highly conductive, it will carry with it the lines of force of the general solar magnetic field. Parker (1963) considered the frozen in magnetic field configuration of interplanetary space. Due to the rotation of the Sun, the field is twisted and is of the form of an archimedian spiral in the Sun's equatorial plane as shown in Fig 3.3(c). Such a field configuration was first inferred from cosmic ray flare observations and has since then confirmed by space craft observations.

Parker (1963) had theoretically estimated that the field at the orbit of the Earth is very nearly radial and is of magnitude 2×10^{-5} gauss. The IMF lines have an average angular velocity equal to the sun's equatorial plane ie the solar wind moves (nearly) radially out ward while the archimedian spiral IMF configuration, on the average, corotates with the sun. The spiralling of the IMF terminates at about 100 AU heliocentric distance. The angle between the archimedian spiral IMF and the helio centric radius vector is called the garden house angleas shown in Fig 3.3(d). The component of the garden house angle is expressed as

$$\tan \psi = Vs / w.r.$$

Where 'Vs' is the solar wind velocity blowing radially out ward 'r' is the helio centric distance and 'w' is the angular velocity of the equatorial region of the sun as seen from the Earth. For a solar wind velocity Vs ~ 400 Km/sec, the angle at the orbit of the Earth (r ~ 1AU) turns out to be ~ 45^0. Parker (1958) has shows that the Archimedian spirals can be predicated by if the magnetic field can be derived from a potential function. Ness and Wilcox (1964) have shown that the direction of IMF is well correlated with average direction of photospheric magnetic field. Svalraard (1972) discovered that the polarity of IMF can be inferred by examining the diurnal variations of magnetometers located near the geomagnetic poles. In about 85% of the time the inferred field obtained by this method indicate the polarity of interplanetary magnetic field. These inferred data are available in SGD and many space missions have also provided observations of the variations of interplanetary magnetic field in both time and helio centric distance from the sun.

North South component (IMF, B_z) :

The change of the interplanetary magnetic field (IMF) is known to be an important factor in causing geomagnetic disturbances. The north-south component of IMF B_z provides an opportunity to make magnetic reconnection between IMF and geomagnetic field. The principal difference between north ward component and south ward component of IMF B_z is the effectivness of coupling to the magnetoshphere. Dungey (1961) first proposed that geomagnetic activity is controlled by the north-south component of the interplanetary magnetic field (IMF) by the process known as magnetic recon-nection. The south ward component of IMF B_z which is anti parallel to the Earth's magnetic field well initial reconnection between the two fields, allowing the efficient transfer of mass, momentary and energy from the solar wind in to magnetosphere. The variation of north-south component of the IMF B_z, when converted in to the geocentric solar magneto spheric (GSM) coordinate system, plays a crucial role in determining the amount of solar wind energy that is transferred to the magneto sphere. (Arnoldy, 1971; Tsurutani and Meng, 1972; Akasofu, 1981; Akasofu et al. 1985; Farrugia et al., 1993). Gold 1962 suggested that the transient B_z component is associated with the of magnetic tongue. The prime cause of geomagnetic storms at Earth are strong dawn to dusk electric field, associated with the passage of south ward directed interplanetary magnetic field, B_z passed the Earth for sufficiently long interval times. Gonzalez et al. (1989) have shown that the energy transfer efficiency is of the order of 10% during intense southward directed IMFs, whereas <1% efficient during intense north ward di-recled IMFs due to viscous interaction and other energy transfer mechanism (Tsurutani et. al., 1992 a).

Disturbance Storm time (D_{st}) index :

The equatorial D_{st} index at a given UT represents magnetic field variations at the dipole equator on the Earth's surface, averaged over local time that are caused mainly by the magneto spheric equatorial currents including the cross tail current. One of the most systematic features of geomagnetic disturbances recorded at low latitudes is the depression, below the quiet time level of the horizontal intensity (Moos, 1910; Sugiura and Chapman; 1960). This depression results from the eastward flowing zonal current system in the magnetosphere called the equatorial ring current. Analysis of magnetic storms clearly reveals that the average disturbance D is a composite of both a universal time part D_{st} and local time dependence part D_s.

$$D = D_{st} + D_s$$

The reference level of D_{st} is such that D_{st} is statistically zero on the days internationally designated as quiet days.

For a uniform distribution of stations in longitude, all located approximately at the same angular distance away from the dipole equator, the mean value of D at any instant of universal time will not contaminated by the local time component D_s. Spatial and/or temporal functions in the ring current intensity are no as rapid as those in the auroral electrojects so that a dense network in longitude is not an absolute necessity. The basic concept of the definition of a D_{st} index invokes magnetic disturbance, in principle the index can be derived as a continuous function of universal time. The definitions of D_{st} and D_s are equally applicable to all three magnetic element, but the magnetic field due to the symmetric zonal current system in the equatorial plane is parallel to the dipole axis and is seen mainly in the H component (for small D_{st} (z) and D_{st} (y) values derived from averages of several magnetic storms.

Results and Discussion

The influence of interplanetary magnetic clouds on cosmic ray intensity and various heliospheric parameters is very complex and hence needs detailed investigations. In the present work we have identified three interplanetary magnetic clouds during the year 2006 to study their effect on cosmic ray intensity variation, geomagnetic field and interplanetary parameters. Using the methodology of Zhang and Burlaga [6] we have identified three interplanetary magnetic cloud events (5, February 2006, 13 April 2006 and 14 April 2006) during 2006 to study their influence on interplanetary, solar wind plasma as well as on cosmic ray intensity.

Figure 1 represents the magnetic cloud event of February 5, 2006. The onset time of magnetic cloud is 1910 Hr UT on day 036. In Fig 1 the upper panel (a) shows the daily average of hourly cosmic ray count rate (Newark neutron monitor) from January 25, 2006 to February 15, 2006 i.e. 11 days prior and 10 days after the onset of magnetic cloud event. The daily average of interplanetary magnetic field (B), north south component of interplanetary magnetic field (Bz), solar wind velocity, sunspot number (R) and disturbance storm time index (Dst) associated with this event is also plotted in different panels of the figure.

As one can see from the plot that the cosmic ray intensity is observed to increase 11 days prior to onset of magnetic cloud event and then decrease for one day and further increases for few days and decreases significantly on February 12, 2006 with some deviations. The interplanetary magnetic field B is increases significantly from 6 days prior to 1 day after the event with some depression on February 4, 2006 then decrease significantly up to 4 days after the event. The systematic variations are seen in the Bz compo nent of interplanetary magnetic field 11 days prior to the event upto the onset of magnetic cloud and then decrease significantly after one day of the event and then increase significantly up to 2 day after the event and then systematic variations continued up to 10 days after the event.

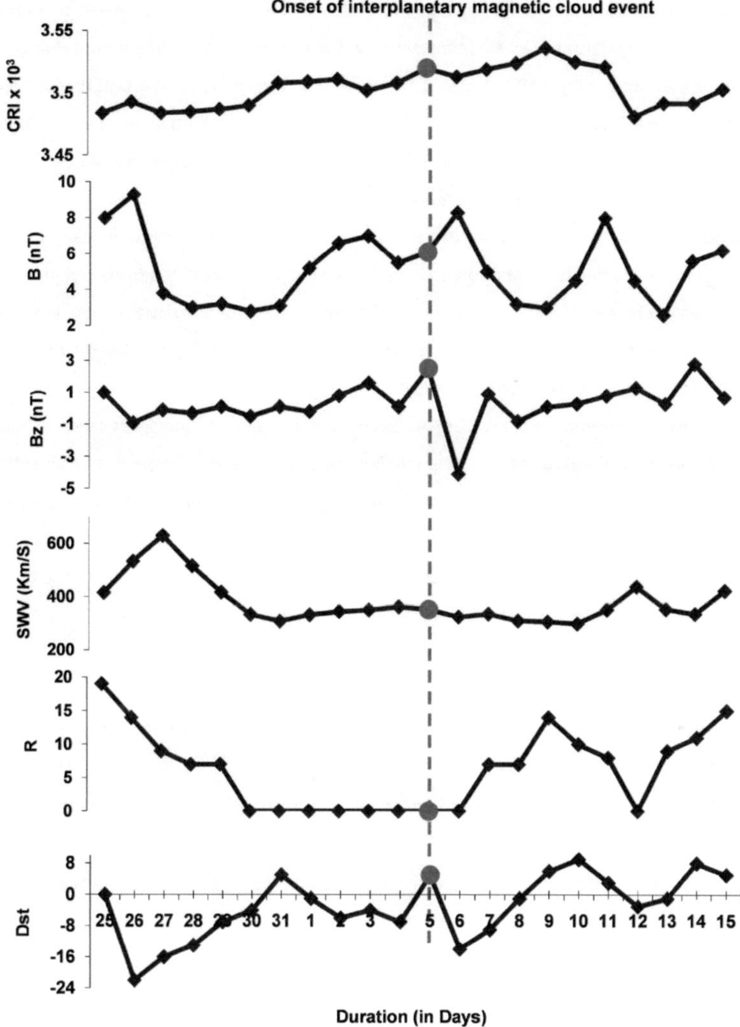

Fig 1: Daily variation of cosmic rays from -11 to +10 day for the magnetic cloud event of February 5, 2006 along with interplanetary magnetic field (B), north-south component of interplanetary magnetic field (Bz), solar wind velocity, sunspot numbers (R), disturbance time index (Dst).

The solar wind velocity is seems to remain constant before and after 5 days of the onset of magnetic cloud. The sunspot number (R) significantly decreases 11 days prior to the event 6 days prior to the event and then significantly remains constant upto 1 day after the event. The abrupt change is observed in the sunspot numbers after 1 day of the onset of magnetic cloud. The disturbance time index, Dst is found to increase significantly 10 days prior to the event up to 5 days prior to the event and then decreases upto 1 day prior to the event. On the onset of magnetic cloud the Dst decreases significantly for 1 day and then increase upto 5 days after the event.

It is noted that on the onset of magnetic cloud the cosmic ray intensity, Bz component of IMF, Disturbance storm time index Dst found to decrease for one day and then all the three components increases gradually, whereas interplanetary magnetic field B increases for one day and then decrease sharply. However the solar wind velocity found to remain constant and sunspot number (R) increases after one day of the onset of cloud with some depression.

The disturbance storm time index (Dst) has been taken as a level of geomagnetic disturbance. North south component of interplanetary magnetic field Bz turns southward immediately after the magnetic cloud event. This southward turning of Bz produces large geomagnetic disturbances, which reflects in Dst value. Increase in Dst index, sunspot number (R) and Bz after the magnetic cloud event seems to be associated with cosmic ray intensity increase. We have rigorously studied all the three magnetic clouds and found the similar trends in all the cases.

Conclusions

From the present investigations following conclusions may be drawn:

- The north south component of IMF (Bz) produce large geomagnetic disturbance on the onset of interplanetary magnetic cloud.
- The deviations in the interplanetary and solar wind plasma parameters are significantly correlated to the magnetic cloud events.
- The increase in Dst index, sunspot number (R) and Bz after the magnetic cloud event produces increase in cosmic ray intensity.

Acknowledgements

The authors are indebted to various experimental groups, in particular, Prof. Margret D. Wilson, Prof. K. Nagashima, Miss. Aoi Inoue and Prof. J. H. King for providing the data.

References

S. P. Duggal, M.A. Pomerantz, R.K. Schaefer, and C. H. Tsao, J. Geophys.

Res., 88, 2973, 1983.

Badruddin, R. S. Yadav, N. R. Yadav, Solar Phys., 105, 413, 1986.

Badruddin, R. S. Yadav, and S. P. Agrawal, Proc. 19th, Int. cosmic ray conf., SH-5, 1-12 NASA Cof. Publ.,
258, 2376, 1985.

S. Kudo, M. Wada, P. Tanskanen, and M. Kodama, 19th, Int. cosmic ray conf. SH5, 1-8 NASA conf. Publ.,
2376, 246, 1985.

N. Iucci, M. Parisi, C. Signorini, M. Storini, and G. Villoresi, 19th. Int cosmic ray conf. Publ., 2376, 226,
1985.

Z. Zhang, and L. F. Burlaga, J. Geophys. Res., 93, 2511, 1988.

L. F. Burlaga, International Magnetohydrodynamics, Oxford Univ. Press, New York, 1995.

L. F. Burlaga, R. Lepping, and J. Jones, Geophys. Monogr. Series, Vol. 58, 373, 1990.

J. T. Gosling, Geophys., Monogram, 58, 343, 1990.

R. P. Lepping, L. F. Burlaga, and J. A. Jones, J. Geophys. Res., 95, 11957, 1990.

C. J. Farrugia, L. F. Burlaga, and R. P. Lepping, Geophys. Monogr. Series. Ed. By Tsurutani et al., 1997

B. T. Tsurutani, and W. D. Gonzalez, The Interplanetary causes of magnetic stroms, a Rev. in Geophys.
Monogr. Series, 1997.

C. J. Farrugia, R. P. Lepping, L. F. Burlaga, A. Szabo, D. Vassiliadis, P. Stauning, and M. P. Freeman, EOS,
Trans. AUG. 77(17), Spring Meet. Suppl., S2 41, 1996.

Burlaga, L.F; 1988, J. Geophys. Res., 93, 7217.

Vandas, M., Fischer, S., Pelant, P. and Geranious, A.; 1993a, J. Geophys. Res., 98, 11467.

Goldstein, H.; 1983, NASA Conf. Publ., 731, 2280.

Burlaga, L.F., Leeping, R. and Jones, J.; 1990, Geophys. Monogr. Series, Vol. 58, 373.

Burlaga, L.F., Sittler, E., Mariani, F. and Schwenn, R.; 1981, J. Geophys. Res., 86, 6673.

Farrugia, C.J., Burlaga, L.F. and Leeping R.P.; 1997, Magnetic Clouds and Quit storm effect at the Earth
in Geophys. Monogr. Series. Ed. By Tsurutani et. al.

Burlaga, L.F.; 1995, International Magnetohydrodynamics, Oxford Univ. Press, New York.

Gosling, J.T.; 1990, Glophys, Monogram, 58, 343.

Farrugia, C.J., Burlaga, L.F. and Leeping R.P.; 1997, Magnetic Clouds and Quit storm effect at the Earth
in Geophys. Monogr. Series. Ed. By Tsurutani et. al.

Farrugia, C.J., Leeping, R.P., Burlaga, L.F., Szabo, A., Vassiliadis, D., Stauning, P. and Freeman, M.P.;
1996, EOS, Trans. AUG. 77(17), Spring Meet. Suppl., S2 41.

Tsurutani, B.T., and Gonzalez,W.D.; 1997, The Interplanetary causes of magnetic stroms, a Rev. in Geophys. Monogr. Series.